The Nature and Science of
FIRE

Jane Burton and Kim Taylor

Gareth Stevens Publishing
A WORLD ALMANAC EDUCATION GROUP COMPANY

Please visit our web site at: www.garethstevens.com
For a free color catalog describing Gareth Stevens Publishing's list of high-quality books
and multimedia programs, call 1-800-542-2595 (USA) or 1-800-461-9120 (Canada).
Gareth Stevens Publishing's Fax: (414) 332-3567.

CHILDRENS ROOM

Library of Congress Cataloging-in-Publication Data

Burton, Jane.
The nature and science of fire / by Jane Burton and Kim Taylor.
p. cm. — (Exploring the science of nature)
Includes bibliographical references and index.
ISBN 0-8368-2198-X (lib. bdg.)
1. Combustion—Juvenile literature. 2. Fire—Juvenile literature.
[1. Fire.] I. Taylor, Kim. II. Title.
QD516.B83   2001
541.3'61—dc21   00-063731

First published in North America in 2001 by
Gareth Stevens Publishing
A World Almanac Education Group Company
330 West Olive Street, Suite 100
Milwaukee, Wisconsin 53212 USA

Gareth Stevens editors: Barbara J. Behm and Heidi Sjostrom
Cover design: Karen Knutson
Editorial assistant: Diane Laska-Swanke

Printed in the United States of America

1 2 3 4 5 6 7 8 9 05 04 03 02 01

Contents

Words that appear in the glossary are printed in **boldface** type the first time they occur in the text.

Burning Bright

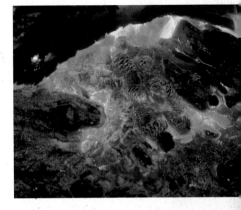

Above: Red-hot charcoal glows for many hours after the flames of a wood fire die down.

When dry branches catch fire in dry weather, they burn very quickly. Flames leap skyward, and a column of smoke rises above the flames. The branches pop, crackle, and glow. The heat is intense. When the fire dies out, only white **ash** and pieces of **charcoal** remain.

Fire consists of heat and light that result when materials burn. Certain factors must be present in order for fire to occur. First, **combustible** materials (also known as fuel) must be present. Second, temperatures high enough to cause **combustion** must be present. This is known as the **kindling point**, or temperatures at which fuel can easily combine with oxygen. Third, enough oxygen to allow rapid combustion must be present. One-fifth of Earth's atmosphere is oxygen.

Most materials that can burn contain plenty of **carbon**. Wood contains a lot of carbon. The lowest temperature at which fire from a carbon and oxygen reaction can continue to burn is about 1,110° Fahrenheit (600° Celsius). Carbon at this temperature glows dull red. At higher temperatures, it glows bright red or yellow. If the temperature of burning wood drops much below 1,110° F (600° C), the fire goes out.

Opposite: A sheet of flame leaps skyward from a grass tree during an Australian bush fire.

Below: White wood ash is all that is left after this fire.

Flames

A wood fire does not just glow red, it also produces bright yellow flames that leap around. Flames occur when **vapor**, or gas, burns. Flames consist of mixed gases undergoing combustion.

The carbon in an unburned log of wood combines with hydrogen and other elements in a wide range of **compounds**. Some of these compounds are **volatile**, which means that they turn into vapor when heated.

During a forest fire, the volatile parts of wood start to turn into vapor before the wood itself becomes hot enough to burn. Flames are created by the combination of this **flammable** vapor with heat and the oxygen in the air.

Right: A fierce fire consumes gorse bushes in a sheet of red flame.

Left: The needles of a young pine tree release flammable vapor when they are heated by fire.

Unburned vapor at the innermost part of a flame is relatively cool. As the vapor combines with oxygen high up in a flame, it glows yellow and is very hot. Above a flame, hot **carbon dioxide** rises into the air.

The temperature in the yellow part of a flame and directly above it may be well over 1,830° F (1,000° C). Blue flames may be even hotter.

Flames from a forest fire may leap many feet (meters) into the air. If there is a wind, the flames are blown around, fueling the fire. The fire spreads quickly to nearby trees.

The twigs and leaves of the neighboring trees become heated. The heat results in more vapor, which produces more flames. In strong winds, fire can spread faster than a person can run. In very little time, hundreds of thousands of acres of mighty forests can quickly be covered by flames.

Below: Vapor oozes from a heated bracken stem and then bursts into flame.

Smoke, Soot, and Ash

Top: Pine cones contain a substance called resin, which vaporizes when heated. The vapor burns with a yellow flame.

When a piece of wood is dipped in liquid oxygen and then lit with a match, it burns instantly. This shows that the rate at which wood burns is controlled by the amount of **free oxygen** available to the fire. Often, wood at the heart of fire is starved of oxygen because the surrounding fire is using it all.

When vapor containing carbon burns without enough oxygen, particles of unburned carbon float off into the air. These particles form smoke. The smoke billowing from a forest fire is not all carbon particles, however. Most of it consists of water droplets formed when the water in living plants is released as *nonflammable* water vapor by the heat of the flames.

A very hot forest fire sends masses of water vapor high into the air. In the cold, upper

Right: Much of the "smoke" from a forest fire is not really smoke at all. It is actually tiny droplets of water. The water in wood and leaves is vaporized by the heat of the fire. The water quickly changes into droplets to produce the billowing, white appearance of smoke.

Left: During a forest fire, warm, moist air is drawn into the cold upper atmosphere. There, the water vapor condenses to form clouds, which may produce rain.

atmosphere, the vapor **condenses** to form clouds. These clouds may produce rain. Surprisingly, a forest fire can eventually result in rain.

Fuel oils, candle wax, and resin formed by pine trees are **hydrocarbons**. All of these substances vaporize easily. When they burn, they form vapor so quickly that there is often not enough oxygen in the surrounding air to burn all of the vapor off. The result is thick, black smoke. This type of smoke collects on solid objects as a soft deposit of **soot**. Soot is pure carbon.

The white, powdery ash that is left after wood burns is made of silica, lime, potash, and other chemicals that contain **combined oxygen**. These substances will not burn.

Left: Charcoal in this wood fire gently smolders to produce fine, white ash.

Fire Lighters

Top: Green leaves contain plenty of water, so fires do not start easily in lush forests.

There has always been enough heat on Earth to start fires. Red-hot rock spewing from volcanoes will set fire to anything that can burn.

In the very early days of our planet, however, there was no free oxygen in the air. Instead, there was carbon dioxide, and so fire was not possible. Millions of years later, bacteria appeared which used energy from the Sun to split carbon dioxide into carbon and oxygen. The bacteria used the carbon in order to grow and released oxygen into the atmosphere. When plants appeared millions of years later, they used even more carbon dioxide and produced even more oxygen.

With oxygen now present in the atmosphere, the environment was ready for fire — and there was no shortage of sparks to light it!

Right: Lava flowing from a volcano sets fire to flammable objects in its path.

There have been forest fires on Earth as long as there have been forests. The main cause of natural fire is lightning. Hundreds of thunderstorms rage over our planet daily. Many storms contain heavy rain, which quickly puts out any fire started by lightning. Often enough, however, lightning strikes a tree on the edge of a storm, and no rain falls to put the fire out. The fire then spreads from tree to tree and may burn for weeks.

When people first discovered how to make a fire, they probably first observed natural sources of fire, such as volcano eruptions and fires started by lightning. They soon learned that fires could make life more comfortable when used for protection, warmth, light, cooking, and farming. These early peoples may also have started some fires accidentally, just as people do today.

The Inferno

Forests in the moist, **temperate** regions of the world may have stood for many centuries without ever being destroyed by fire. These forests are too damp for fire to start easily, and few thunderstorms occur to provide the sparks. When fire does blaze up in these areas, however, the plant life is unable to resist it.

Forest fires in temperate regions are usually started by people and are very destructive. Leaves become scorched and burst into flames in seconds as fire sweeps through the treetops. Small branches may be consumed by the fire, while the main branches and trunk of a tree are left standing. Yet, even those solid woody parts of the tree may not survive. Heat from the fire can penetrate through the bark and kill the growing layer just beneath. Once this has happened, the tree is doomed. It will stand, black and stark, for years after the fire. Its place will eventually be taken by new trees.

Fire can spell death for many animals, too. Insects and other invertebrates (animals without backbones) cannot move fast enough to escape an advancing forest fire and are burned alive. Many small vertebrates (animals with backbones), such as lizards, frogs, and mice, are trapped by fires. Even larger animals, such as deer, may lose their sense of direction in the smoke and be caught by the flames.

Top: In a forest fire, mice may run for their lives or hide in holes underground.

Opposite: Fires do not occur often in the damp forests of temperate regions. When there is a fire, many of the trees there cannot survive the heat. These tall lodgepole pines were killed by a fire that swept through four years earlier.

Below: The bark of this ponderosa pine burned away in a forest fire. No tree can survive the loss of its bark.

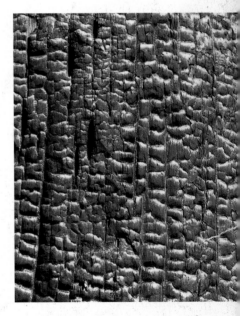

Fire Survivors

Top: **Top:** The heat of fire and the dampness of rain caused this firewood banksia cone to open and spill its seeds.

Above: The bull banksia produces a hard, woody cone that can survive fire.

Right: The bark of this giant sequoia in Oregon is thick and spongy. It protected the tree during the many fires that have swept through the area over the past 1,500 years.

In warm parts of the world, fires that begin naturally occur every few years. Plants in these areas have developed several different ways of coping with fire. Some trees grow thick, spongy bark. This type of bark protects the delicate growing layer within.

Cork, from which bottle corks are made, is the bark of the cork oak tree. Cork does not easily catch fire. When it does, it **smolders** for a while and then goes out. Cork is a very bad **conductor** of heat. Even if the outside of a cork oak tree is smoldering, the heat does not get through to the wood underneath. Fire may kill the leaves and branches of a cork oak, but the tree lives. After the

Left: This banksia bush in Western Australia was killed by a fire. Although the tree is dead, the hard cones on the ends of the branches protect the seeds.

fire passes, the tree will eventually sprout new branches and leaves. Cork oaks grow in dry areas where fires are common. They have developed their very own fire protection system.

During a fire, small plants and bushes are usually killed, but some of them leave plenty of seeds behind. Large, hard cones surround the seeds of banksia bushes in Australia. For years after a fire, the cones may remain on the bushes with the seeds locked inside, waiting for another fire.

When fire comes, the cones are scorched and begin to dry out. As the cones dry, they open and release the winged seeds inside. The wind blows the seeds away, and they fall to the ground and **germinate** as soon as the winter rains arrive.

In parts of the world where fires regularly occur, many of the plants produce fire-resistant cones.

Below: The seed pod of an Australian hakea bush burned and split open. The outside of the pod is charred, but the winged seeds inside are alive. They are embedded in thick, corky material.

Right: The heat of an Australian bush fire is intense. The above-ground parts of low-growing plants die, but many trees survive.

The seeds of some plants that grow where fires are common cannot germinate *until* they have been exposed to fire. Chemicals in the smoke and ash are enough to make the seeds germinate, so the seeds do not necessarily have to be heated by the fire in order to grow. The first plants to grow in a burned area have a good start in life. There are no other plants to compete with them, and ash from the fire provides a layer of fertilizer.

Right: The roots of this cowslip orchid survived a bush fire. Its flowers are one of the first signs of renewed life to show above the fire-blackened soil.

Left: A lone tammar wallaby views the desolation of its homeland after a bush fire.

Above: Two weeks after a bush fire in Australia, green shoots spring from the blackened trunks of some of the trees.

Unlike plants, animals can escape fire by moving away from it — and many do. Birds fly to safety, and the larger land animals have enough intelligence to know how to escape being burned. The noise and smell of a fire warn them to keep moving ahead of the flames.

Not all animals run away from fires, however. In the Australian bush, where fires have occurred regularly for thousands of years, some animals behave unusually.

The small, kangaroo-like tammar wallabies that live in the dry forests of southwestern Australia may run back and forth *alongside* an advancing wall of flames. As they run, they look for a spot where the fire is less intense. There, they rush through the flames, with scarcely singed feet, onto the burned land beyond. The wallabies will be short of food for a week or two until the bushes start to sprout again. The main thing, however, is that they have survived the fire that swept their home.

Above: Six months later, flowering plants grow.

Below: Eighteen months after the fire, vegetation covers the ground.

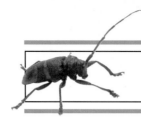

Fire Lovers

Top: The **larvae** of this longhorn beetle in southern Africa feed on dead wood. Trees killed in a fire provide the wood.

Fire in the natural world is not always bad for plants and animals. Fires burn away tough, dead grass, leaving the roots undamaged. Soon, fresh green grass grows, providing nutrition for grazing animals.

Some ranchers even start fires on grasslands so that their cattle will have fresh grass to eat. Natural fires, which sweep through the grasslands of Africa from time to time, provide fresh grazing for the herds of antelope that live there.

A grass or bush fire is also good for many **predators** and **scavengers**. Certain birds, such as kites, eagles, and vultures, often fly to smoky areas. There, they find small animals running or flying for their lives from a fire. The predatory birds swoop down to catch and eat grasshoppers, lizards, and other small creatures. After the fire dies down, scavengers feed on the animals that died in the fire.

Below: Waterbuck graze on fresh grass that has sprung up after a fire in Africa.

Left: Black kites swoop over a fire in East Africa. These birds see smoke from far away and fly to the fire. On the ground, small animals try to escape the heat and flames. The kites fly to the ground and use their talons to grab the animals.

A forest fire in temperate regions results in an abundance of dead wood. The dead wood provides food for various kinds of insect larvae. There is just one problem. If the fire has swept through a large enough region, all the insects in the area may have died in the fire. There may be no insects left to lay eggs for a new generation.

The blue beetle (*Melanophila acuminata*) has solved this problem by developing heat sensors within its body. Remarkably, the sensors can detect forest fires that are burning up to and over 30 miles (50 kilometers) away. The sensors guide the beetles toward the fire. The beetles fly into the burning area as the fire is dying down. Then they lay their eggs on blackened, smoldering branches.

Larvae of the blue beetle and the Buprestidae family of beetles get the best of the dead wood to eat because their parents got to the fire first.

Below: An adult member of the Buprestidae family of beetles (metallic wood-borers) flies toward a forest fire as the fire is dying down and lays eggs in the dead trees. The larvae feed on dead wood.

Right: Fires on the moorlands burn the old, woody heather plants, making way for fresh, green shoots to grow. Red grouse, like this one, feed on the young heather.

Fire is a positive thing for some animals in the heather **moorlands** of temperate regions. Heather is a woody plant that grows very slowly, reaching a little over 3 feet (1 m) in height. For the most part, mature heather plants are tough and poor in food value. Nevertheless, some animals survive by eating only heather plants.

Red grouse and mountain hares rely on heather for food. The *young* heather plants provide these animals with the best quality food. Heather quickly sprouts new growth after mature plants burn. More grouse and hares live in areas where moors are burned on purpose than in areas where the moors are left alone.

Below: A tame rook named Niger hammers the tip of a match with his beak, trying to light it.

Below: The match starts on fire, and Niger reacts by spreading his wings over the flame.

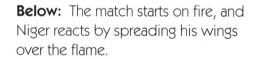

Left: Mountain hares eat the fresh green heather shoots that appear after a fire.

Some species of birds have been observed cleaning themselves with the acid produced by ants. These birds pick up live ants and rub them quickly over their feathers to kill other insects or treat their skins. This activity is called "anting."

It is possible that some birds have actually learned to use fire. Niger was the name of a rook, which is similar to a crow. Niger became very excited whenever a small fire was started for him. He would wallow in the flames until his feathers became singed and he had put the fire out. He even learned to light a match by hammering it with his beak. He would then rub the match under one wing, as if "anting."

Below: Niger flaps his wings over a small pile of straw that is on fire, perhaps cleaning himself with the flames.

Below: Niger picks up a piece of smoldering straw and rubs it over his feathers as though "anting."

21

Fire Inside

Top: Herbivores, like this guinea pig, have to eat all day long. Plants are a low-grade fuel.

Above: Seed-eaters, like this red squirrel, receive plentiful energy from each seed or nut they eat. Nuts contain oil and **carbohydrates**, which are energy-rich foods.

When a fire burns, fuel and oxygen combine to produce heat, and heat is a form of energy. Similarly, food is fuel for animals. When animals digest a meal, energy is produced. In fact, digestion is like a controlled fire. Oxygen, breathed in by the animal, mixes with fuel (the food) to produce carbon dioxide and heat. The animal breathes out the carbon dioxide and uses the heat for energy.

The amount of energy and warmth that an animal can get from food depends on the type of food the animal eats. The grass and leaves that **herbivores** eat contain mostly **cellulose** and water. Cellulose is mostly made up of carbon that is difficult for animals to digest. Herbivores spend much of their lives eating and digesting in order to obtain enough energy to survive.

Right: After this lioness fills her stomach with zebra meat, she may not need to eat again for three days. The protein in the meat is a high-grade fuel.

Seeds and nuts contain easily digestible carbohydrates, which are made up of carbon, hydrogen, and oxygen. Carbohydrates are rich in energy, so seed-eaters can quickly obtain energy from their food. The meat that **carnivores** eat contains mostly protein, which is also rich in energy. A lion gets enough energy from one good meal of meat to last for three days.

Hydrocarbons, which contain carbon and hydrogen only, burn easily. An example of a hydrocarbon is candle wax. Fat and oil are made up mainly of carbon and hydrogen, too, so processed animal fat, called tallow, also burns well. Animal fat and the oils from plant seeds are the foods richest in energy. They "burn" in the body to produce large quantities of heat.

Above: The acorns that fall from oak trees in autumn provide animals with a source of energy-rich food. The energy in an acorn can also provide a good start for a young oak tree.

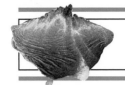

Fire Below

Top: Molten lava glows brightly as it flows out of a volcano.

A fiery plume spurts from the top of a volcano. The mountain looks like it is on fire. The heat in a volcano is not caused by fire, however. Nothing is burning. What looks like fire is, in fact, red-hot **molten** rock. The heat that melts the rock comes from deep inside the Earth, where there is no free oxygen — so fire is impossible.

The heat within the Earth is generated by **nuclear energy**. Heavy forms of elements, such as uranium, slowly change to lighter forms in a process called **radioactive decay**. This process releases huge amounts of energy as heat. Radioactive decay has been going on inside the Earth ever since the planet formed over four billion

Right: Red-hot lava pours over a low cliff and forms a pool at the bottom. The heat comes from deep inside the Earth.

Left: As lava cools, a crust forms on top of it. No ash is left behind because nothing burned.

Below: Cracks in cold lava soon provide a foothold for plants. The plants grow well because lava contains nutrients.

years ago. The final product of radioactive decay may be iron. The central core of Earth is believed to be a massive pool of red-hot, liquid iron. This pool of iron creates the magnetic field that surrounds our planet.

When a volcano bursts with a fiery eruption, liquid rock, called lava, pours out. As the lava flows down the side of the mountain, it slowly cools and becomes solid. Red-hot, moving lava will set fire to any flammable object near it. For this reason, forests on the sides of active volcanoes are frequently ablaze with fire.

Left: Red-hot lava pours over a road, setting it on fire.

The Burning Sun

Top: The fiery glow of the setting Sun is not caused by fire. The Sun's heat is produced by nuclear energy.

The temperature in the Sun's **convection zone** is about 3,600,000° F (2,000,000° C). The Sun's matter is white-hot, but the Sun does not burn because there is no oxygen to support fire. The Sun's heat, like Earth's heat, comes from nuclear energy. The nuclear reaction in the Sun is very different from the nuclear reaction in Earth. There is no radioactive decay — only **nuclear fusion**.

The Sun consists of a huge ball of white-hot gases. Much of the gas is hydrogen, but there is also a lot of helium. Helium forms when two atoms of hydrogen join together, or fuse. When nuclear fusion takes place, energy is released. The high temperature and huge pressure inside the Sun cause hydrogen atoms to fuse continuously. This process has been going on for billions of years and will continue for billions more. The Sun's nuclear fuel will run out one day, causing our Sun to dim — but it won't happen for a very long time!

Right: The Sun is a giant ball of white-hot gas. The huge amount of heat released by the Sun day after day, year after year, is produced by nuclear reactions inside of it.

Without heat and light from the Sun, there would be no life on Earth. Plants use the Sun's energy to take carbon out of the atmosphere and return oxygen. Fire is a product of plant life.

A burning forest reverses the process of storing the Sun's energy. In a forest fire, oxygen combines with carbon. The Sun's energy that was taken in years before is released as heat. These processes have been going on as long as there have been plants on Earth.

Fire can be a destructive force, but it is also a positive force in nature, allowing new plant and animal life to grow.

Above: When a forest burns, the Sun's energy — which has been collected and stored by the trees — is released as heat. Beneath the charred trunks, flowering plants such as the rosebay willowherb or the hawksbeard spring to life, fertilized by chemicals in the ashes.

Activities:
Controlling Fire

When carbon burns, it combines with free oxygen in the air to produce heat and carbon dioxide. To see how providing extra oxygen causes charcoal to glow even more brightly, gently blow on a piece of glowing charcoal.

Given enough oxygen, the charcoal will eventually completely burn out, leaving only ash. Without oxygen, charcoal can be heated until it is red-hot, but it *will not burn*. It just stays red-hot.

In many parts of the world, charcoal is used as a fuel for cooking. A charcoal fire produces heat without smoke and can be controlled by fanning or blowing on it to raise its temperature. Charcoal furnaces with bellows that produce a strong draft of air were used in the past to melt iron at over 2,190° F (1,200° C).

Homemade Charcoal

To make your own charcoal, you will need a tin can with a tight-fitting lid, a hammer, a large nail, and some small pieces of wood or dead twigs from a bush or tree. If you cannot find a can with a lid, use two lidless cans that fit over one another (*top right*). You will also need a fire that contains a bed of glowing embers. Do this experiment outside — in the remains of a fire that was used to burn dry branches, for instance (*bottom right*).

Before you start, ask an adult to help you with this experiment — especially to make sure the fire is safe.

With the hammer and nail, punch a small hole in the lid of the can. Next, load the can with wood pieces or twigs. Put the lid on firmly. **With adult supervision,** place the can more or less upright in a hot part of the fire. Next, watch and wait.

As the can heats up, volatile substances in the wood vaporize and are forced out though the hole in the lid. Water is first to come out as steam. As the can becomes hotter, flammable vapor jets out of the hole and burns with a yellow flame (*next page, top*). The vapor cannot burn inside the can because there is not enough oxygen there. For the same reason, the carbon in the wood remains unburned.

When the jet of flame from the can finally goes out, carefully remove the can from the fire. (*This is best done with long tongs, a pitchfork, or a shovel.*) The can will be **very hot**, so leave it for a long time before touching it. Make sure the lid stays on the can. Otherwise, the charcoal may start to burn. After the can cools, open it. Inside will be black sticks of charcoal.

Use the charcoal to make drawings on sheets of paper. Also try lighting one of the pieces with a match, blowing on it to keep the red-hot ember glowing.

The Disappearing Flame

Candle wax is a hydrocarbon. It is a compound of hydrogen and carbon that burns quickly and easily. When you light the wick of a candle (*bottom right*), the heat of the match first melts the wax and then causes it to vaporize.

The wax vapor burns with a yellow flame, drawing in oxygen from the air in order to do so.

To show that oxygen is actually used by the flame, you will need a candle (*just like the one pictured, top right*), a large glass jar, a shallow dish partly filled with water, a lid from a tin can, and a match.

Place the candle on the lid from the tin can. Float the lid and the candle on the water in the dish. Light the wick of the candle. Gently lower the jar (*upside down as pictured below*) over the burning candle so that it stands in the water.

The flame now has only the oxygen that is within the jar to keep it burning.

As the oxygen is used up, the flame becomes weaker and weaker until it finally goes out.

As the flame dies out, notice that the level of the water in the jar has risen (*above, right*).

The reason for the rise in the level of water is due to the presence of hydrogen. When hydrogen burns, it forms water vapor. As the vapor cools, it quickly returns to being liquid water.

Liquid water occupies far less space than the oxygen gas that was used in its formation. So some of the water from the dish is drawn up into the jar.

Glossary

ash: material, such as minerals, that does not burn and remains after a fire.

carbohydrates: chemicals formed of carbon, hydrogen, and oxygen.

carbon: an element that can be found in coal, diamonds, and all living things.

carbon dioxide: a gas, exhaled by humans, combining carbon and oxygen.

carnivores: meat-eaters.

cellulose: a tough carbohydrate found in nearly all plants.

charcoal: the substance remaining after the volatile parts of wood are driven off by heat. Charcoal is mostly carbon.

combined oxygen: ordinary oxygen with two oxygen atoms that is combined with another chemical element to form a new substance.

combustible: able to burn.

combustion: the rapid chemical process that produces heat and usually light; fire.

compounds: substances made up of more than one element.

condenses: changes from vapor to liquid.

conductor: a material that can transmit, or pass on, a form of energy, such as heat.

convection zone: the outer third of the Sun's interior, ending just below the Sun's surface.

flammable: able to catch fire.

germinate: to start to grow.

herbivores: plant-eaters.

hydrocarbons: chemicals that contain only hydrogen and carbon.

kindling point: temperature at which fuel can easily combine with oxygen.

larvae: the young, wingless stage in the growth of insects. *Singular: larva.*

molten: melted; liquefied by heat.

moorlands: open, rolling, infertile land.

nuclear energy: energy released when atoms split or form.

nuclear fusion: the process by which light elements (such as hydrogen) combine to produce heavier elements and release energy.

predators: animals that hunt other animals for food.

radioactive decay: the natural process by which certain forms of elements give off energy and change into other forms of the same elements or into other, lighter, elements.

scavengers: animals that feed on the remains of dead animals.

smolders: burns slowly, without flame, but with a lot of smoke.

soot: finely-powdered carbon.

temperate: moderate; not very hot or cold.

vapor: a gas formed from a liquid.

volatile: able to evaporate readily.

Plants and Animals

The common names of plants and animals vary from language to language. Their scientific names, based on Greek or Latin words, are the same the world over. Each kind of plant or animal has two scientific names. The first name is called the genus. It starts with a capital letter. The second name is the species name. It starts with a small letter.

African lion (*Panthera leo*) — Africa 22

black kite (*Milvus migrans*) — Africa, Europe, Asia 19

bracken (*Pteridium aquilinum*) — worldwide 7

bull banksia (*Banksia grandis*) — Australia 14

Buprestidae beetle (*Agrilus pannonicus*) — Europe 19

common waterbuck (*Kobus ellipsiprymnus*) — Africa 18

common zebra (*Equus burchelli*) — Africa 11

cowslip orchid (*Caladenia flava*) — Australia 16

giant sequoia (*Sequoiadendron giganteum*) — North America 11, 14

gorse (*Ulex europaeus*) — Europe 6

grass tree (*Xanthorrhoea preissii*) — Australia 4–5

guinea pig (*Cavia porcellus*) — domesticated 22

hakea (*Hakea crassifolia*) — Australia 15

hawksbeard (*Crepis* species) — North America 27

lodgepole pine (*Pinus contorta*) — North America 13

mountain hare (*Lepus timidus*) — northern Europe 21

pedunculate oak (*Quercus robur*) — Europe 23

red grouse (*Lagopus lagopus*) — northern Britain 20

rook (*Corvus frugilegus*) — Europe 20–21

rosebay willowherb (*Epilobium angustifolium*) — Europe, North America 27

tammar wallaby (*Macropus eugenii*) — Australia 17

Books to Read

Fire Line: The Summer Battles of the West. Michael Thoele (Fulcrumb)

Fire in Their Eyes: Wildfires and the People Who Fight Them. Karen Magnuson Beil (Harcourt Brace)

How Animals Protect Themselves. Animal Survival (series). Michel Barré (Gareth Stevens)

The Sun and Its Secrets. Isaac Asimov's New Library of the Universe (series). Isaac Asimov (Gareth Stevens)

Volcanoes: Fire from Below. Wonderworks of Nature (series). Jenny Wood (Gareth Stevens)

World Fire. Stephen J. Pyne (University of Washington Press)

Videos and Web Sites

Videos

America's Most Dangerous Volcanoes. (Simitar)

Born of Fire. (National Geographic)

California Firestorm. (American Home Entertainment)

The Expanding Universe: The Sun and Other Stars. (World Almanac Video)

Raging Planet: Fire. (Discovery)

Web Sites

www.wildfirenews.com

www.solarviews.com/eng/sun.htm

whyfiles.org/018forest_fire/index.html

ippex.pppl.gov/ippex/

pearl1.lanl.gov/periodic/elements/8.html

www.uky.edu/Projects/Chemcomics/html/oxygen.html

www.learner.org/exhibits/volcanoes/

Some web sites stay current longer than others. For further web sites, use your search engines to locate the following topics: *combustion, fire, kindling point, nuclear energy, smoke,* and *Sun.*

Index